The Brain–Immune Connection

Training Your Mind to Prevent Chronic Diseases

By Joe Correa

Published by Live Stronger Faster

This book is for educational and informational purposes only. The author and publisher are not engaged in rendering legal, financial, or other professional advice. Readers should consult a qualified professional before making any financial decisions.

First Edition

Table of Contents

Introduction

The human body is a masterwork of integration. Every system—neurological, cardiovascular, immune, endocrine, and digestive—communicates through a constant stream of biochemical messages, electrical impulses, and subtle energetic exchanges. Yet for most of modern medical history, we've been taught to think of the body as a machine made up of separate parts: a brain that thinks, an immune system that defends, a heart that pumps. In truth, these systems are inseparable. What affects one, affects all. And perhaps nowhere is this more profound than in the connection between the brain and the immune system.

In the last two decades, scientific discoveries have radically changed how we understand health and disease. The emerging field of psychoneuroimmunology—the study of how the mind, brain, and immune system interact—has shown that our thoughts, emotions, and stress levels have measurable effects on immune function. Chronic stress, unresolved trauma, and negative thought patterns don't just make us feel bad; they alter our brain chemistry, suppress our immunity, and create the biological conditions for chronic illness. Conversely, practices that regulate the mind—such as mindfulness, visualization, gratitude, and meditation—can enhance immune resilience, reduce inflammation, and even reverse disease processes.

At the center of this relationship is communication. The brain and immune system constantly "talk" through chemical messengers called cytokines, neurotransmitters, and hormones. When the brain perceives a threat—real or imagined—it signals the release of stress hormones like cortisol and adrenaline. In small doses, this helps us survive. But when the stress response is activated too often, it becomes toxic. Chronic elevation of these stress chemicals suppresses immune function, disrupts sleep, promotes insulin resistance, and damages the very neural networks responsible for emotional regulation and memory. Over time, this biochemical imbalance contributes to diseases such as diabetes, autoimmune disorders, heart disease, and even certain cancers.

The opposite is also true. When we cultivate positive mental states—calm, joy, focus, love—the brain releases neurotransmitters such as dopamine, serotonin, and oxytocin. These "healing chemicals" support immune function, improve circulation, and help the body repair itself at the cellular level. This is why a positive mindset is not simply an attitude—it's a biological intervention. The science now shows that mental training is as essential to long-term health as diet, exercise, and sleep.

Imagine your immune system as an army and your brain as the general. The general's job is to interpret information and decide whether to attack, rest, or rebuild. But if the general is constantly overwhelmed by fear, anger, or exhaustion, the army becomes confused. Sometimes it attacks the wrong targets (as in autoimmune diseases), and sometimes it becomes too passive, allowing infections or tumors to grow unchecked. By training your mind to remain calm, clear, and balanced—even in the face of adversity—you teach your "general" to make better decisions. Over time, this creates a state of internal harmony that prevents disease before it starts.

One of the most remarkable aspects of the brain–immune connection is its plasticity. The brain is not fixed; it changes based on how you use it. Neural pathways that fire together, wire together. This means that every time you choose a thought, you are literally training your brain to influence your immune system differently. When you practice gratitude daily, focus on healing images, or use breathwork to regulate your nervous system, you are building new neural networks that signal safety to your body. The immune system listens to this message and adjusts its behavior accordingly. It begins to conserve resources, reduce inflammation, and activate repair processes. Healing becomes not just possible, but predictable.

The implications are profound. Modern medicine has often focused on treating symptoms and suppressing inflammation rather than addressing the deeper causes of immune dysfunction. But chronic inflammation—what scientists now call "the silent killer"—is often the result of unregulated mental stress. The body's internal alarm system never turns off. It's like living next to a fire alarm that won't stop ringing; eventually, the entire system burns out. Reconnecting with the mind and learning to calm the inner storm is one of the most effective ways to turn off that alarm and allow the body to heal naturally.

This book is about reclaiming that power. It's not about replacing doctors or ignoring medical science. It's about complementing traditional approaches with mental and emotional training that strengthens your immune system from the inside out. You will learn how thoughts and emotions directly influence your biology, how to detect early signs of imbalance, and how to use proven mental exercises to restore harmony. You'll explore powerful tools such as neuroplasticity training, breath control, emotional release, and focused visualization—all backed by emerging neuroscience and immunology.

Throughout these pages, you will also encounter real stories of transformation: individuals who overcame autoimmune disorders, chronic fatigue, or even cancer by retraining their minds and restoring communication between the brain and the body. Their recoveries are not miracles—they are the result of conscious rewiring of neural and immune circuits. These examples demonstrate what is possible when we understand the mind not as a passive observer, but as an active participant in our healing.

You are not a victim of your biology. You are its designer. Every thought, emotion, and intention you hold is either strengthening or weakening your immune defenses. Once you recognize this truth, you can no longer separate mental health from physical health. The same mental patterns that cause anxiety and depression can also trigger inflammation and immune dysfunction. The same inner peace that promotes emotional well-being also activates healing at the cellular level. Your health, therefore, begins in the mind.

The journey you are about to begin is not about perfection; it's about awareness. It's about understanding the language your body speaks and learning to communicate back in a way that promotes balance, repair, and longevity. When you train your mind, you train your immune system. When you heal your thoughts, you heal your cells. The brain–immune connection is the bridge between science and spirit, between knowledge and healing—and once you learn to cross it, your understanding of health will change forever.

Chapter 1: The Hidden Dialogue Between Mind and Immunity

The immune system is often described as the body's defense force—a vast network of cells, tissues, and organs working tirelessly to protect us from infections, toxins, and cellular damage. But few realize that the commander of this army sits not in the thymus or lymph nodes, but in the brain. The brain and immune system are in constant communication, exchanging billions of messages per second through a complex network of signaling molecules. This dialogue determines whether the body mounts a defense, rests, or repairs itself.

For decades, scientists believed the brain and immune system were separate entities. The brain was thought to be "immune-privileged," protected by the blood-brain barrier from immune activity. But this idea has been overturned. We now know that immune cells patrol the brain's outer layers, and that neurons and immune cells share similar receptors for hormones and neurotransmitters. The brain sends chemical messages—like adrenaline and cortisol—that influence how immune cells behave. In return, immune cells send cytokines back to the brain, shaping how we feel, think, and respond to stress. This means that every emotional state, from joy to anxiety, leaves a traceable imprint on immune activity.

The brain's interpretation of safety or danger is central to this dialogue. When the brain senses threat—whether physical, emotional, or imagined—it activates the hypothalamic-pituitary-adrenal (HPA) axis. This triggers the release of stress hormones that prepare the body to fight or flee. In short bursts, this reaction is lifesaving. But when prolonged, it becomes damaging. Chronic activation of the stress response leads to sustained cortisol release, which weakens immune cells, shrinks the thymus, and promotes inflammation throughout the body. This low-grade, systemic inflammation is at the root of most chronic diseases, including heart disease, diabetes, autoimmune disorders, and even neurodegenerative conditions like Alzheimer's.

Understanding this hidden dialogue allows us to influence it consciously. The mind can regulate immune activity by altering perception. For example, studies on mindfulness meditation have shown that consistent practice reduces C-reactive protein (CRP) and other inflammatory markers in the blood. Similarly, visualization techniques have been shown to enhance antibody production and increase natural killer (NK) cell activity—cells responsible for destroying infected or cancerous cells. In these cases, mental imagery is not wishful thinking; it is biological training. The brain cannot distinguish between vividly imagined experiences and real ones, and it sends corresponding chemical signals to the body regardless.

Neuroplasticity plays a central role in reshaping this communication. Every time we respond calmly to stress instead of reacting with fear or anger, we rewire neural circuits that govern the body's response to threat. This retraining gradually restores balance to the autonomic nervous system—the division responsible for regulating unconscious processes like heart rate, digestion, and immune activity. When the parasympathetic (rest-and-repair) branch becomes dominant, the body enters a healing state. Cells regenerate, inflammation subsides, and immune responses become more intelligent and efficient.

To harness this power, one must first develop awareness of mental and emotional states. Most people live in a constant state of background stress without realizing it. Their breathing remains shallow, their thoughts race, and their bodies are perpetually braced for conflict. This mental tension translates directly into immune dysregulation. Techniques such as diaphragmatic breathing, meditation, journaling, and mindful observation teach the brain to detect and release unnecessary tension. With time, this practice becomes a subconscious habit—one that maintains the internal environment necessary for health and repair.

Nutrition, sleep, and exercise also shape the brain–immune dialogue. Omega-3 fatty acids, antioxidants, and plant-based nutrients reduce neuroinflammation and support healthy neurotransmitter function. Regular physical activity stimulates immune circulation, while quality sleep allows the brain to clear toxins and reset immune signaling. Yet the most underappreciated factor remains the way we think. A body nourished by healthy food but poisoned by chronic stress cannot thrive.

In this chapter, remember that your mind is not a passenger in the vehicle of your body—it is the driver. By becoming aware of how thoughts and emotions affect immune signaling, you gain the ability to influence your body's health from within. The conversation between your brain and immune system is happening whether you are conscious of it or not. The question is whether you will let it continue by default—or take control of the dialogue and direct it toward healing.

When you learn to regulate the mind, you begin to regulate the immune system. The skill is not mystical; it is biological literacy. Through simple daily mental training, you can create measurable improvements in immune strength, energy, and resilience. The first step is awareness—the ability to observe your internal state without judgment. From there, the path leads to mastery, where your mind becomes your most powerful medicine.

Clara's symptoms

It began with fatigue—an exhaustion that sleep couldn't cure. For months, Clara had been pushing herself harder at work, ignoring the growing ache in her chest and the persistent colds that came more frequently than before. When her doctor ran the usual tests, the results came back "normal." There was no infection, no major deficiency, no sign of disease. Yet Clara knew something inside her was failing.

Her mornings were filled with tension before she even opened her eyes. She would wake up already anticipating the flood of emails and the expectations waiting for her at the office. Her breath was shallow, her heart often racing. She dismissed it as stress—something everyone had. But what she didn't realize was that her body was interpreting every anxious thought as a threat. Her brain was signaling her immune system to stay on high alert, flooding her bloodstream with cortisol and inflammatory cytokines day after day.

One evening, after missing yet another dinner with her family, Clara found herself scrolling through an article titled *"The Brain–Immune Connection."* It claimed that chronic stress could suppress immunity and trigger inflammation even without infection. She was skeptical but intrigued. The next day, she booked a consultation with a neuroscientist-turned-health coach who specialized in psychoneuroimmunology.

During their first session, the coach explained how the mind and immune system were constantly talking. "Your body doesn't know the difference between real danger and imagined danger," he said. "If your thoughts are filled with fear or frustration, your brain sends out distress signals as if you were under attack."

He asked Clara to close her eyes and take a deep breath. At first, it felt forced. But after several slow inhales and long exhales, something shifted. Her shoulders dropped slightly, and her pulse slowed. "You just changed your immune activity," he told her. "You've shifted your body from a defensive state to a repair state. This is how healing begins."

Over the next few weeks, Clara practiced mindfulness every morning for fifteen minutes before work. At first, her thoughts wandered endlessly. But gradually, she began to notice patterns—how certain worries triggered tightness in her stomach, how deadlines made her chest feel heavy. She learned to name these sensations and breathe through them. Instead of resisting her emotions, she observed them.

As the months passed, subtle changes appeared. Her chronic colds disappeared. The skin irritation that had bothered her for years faded. Her sleep improved. But what surprised her most was the clarity of her thinking. For the first time in years, she felt calm enough to make decisions without panic. She began to understand that her immune system had not been broken—it had been overwhelmed by mixed signals from her own mind.

Clara's doctor was puzzled by her transformation. "Your labs look better than they ever have," he said. "Whatever you're doing, keep doing it." She smiled, knowing that her medicine came not from a bottle, but from within. Her coach had helped her realize that the brain–immune connection was not an abstract theory—it was a living dialogue she could influence.

Years later, Clara began teaching stress-reduction workshops to other professionals. She told her story not as a miracle, but as evidence of what happens when the brain learns to speak kindly to the body. Her message was simple: healing begins when we stop fighting ourselves.

Today, Clara still meditates every morning, not because she's sick, but because she knows it keeps her immune system strong. She has come to see her thoughts as nutrients—each one capable of nourishing or depleting her health. Through awareness and practice, she discovered that the most powerful form of immunity is inner peace.

In understanding this, Clara became her own healer. She learned that the hidden dialogue between mind and immunity is not silent—it's just subtle. And when you finally learn to listen, your body begins to speak back in the language of balance, vitality, and lasting health.

Chapter 2: Stress, Inflammation, and the Immune System's Breaking Point

Stress is not inherently harmful. In small doses, it keeps us alert, motivated, and capable of responding to life's challenges. But when stress becomes chronic—when the brain interprets everyday situations as ongoing threats—the body's defense systems begin to malfunction. The same mechanisms that protect us in emergencies start to destroy us slowly from within.

At the center of this process lies inflammation. Inflammation is the body's natural healing response. When tissue is injured or an infection occurs, immune cells rush to the site, releasing signaling molecules called cytokines to contain the damage. For a short period, this is beneficial. But when inflammation becomes chronic, it stops being protective and starts being destructive. Chronic stress keeps the immune system "switched on," flooding the body with pro-inflammatory chemicals even when there is no real danger.

The brain orchestrates this through the hypothalamic-pituitary-adrenal (HPA) axis and the sympathetic nervous system. When you perceive a threat, your hypothalamus signals the adrenal glands to release cortisol and adrenaline. These hormones temporarily suppress non-essential functions such as digestion, reproduction, and even immune regulation—so that your body can focus on survival. Once the threat passes, cortisol levels are supposed to drop. But if the stress never ends, the feedback loop breaks down. The body remains stuck in a low-grade fight-or-flight mode, and the immune system becomes confused—alternating between suppression and overactivation.

This imbalance leads to a paradoxical state where the body is both inflamed and immunocompromised. You might get sick more easily, recover more slowly, or develop conditions linked to chronic inflammation such as arthritis, asthma, irritable bowel syndrome, or autoimmune diseases. In the brain, prolonged inflammation disrupts neurotransmitter balance, contributing to anxiety, depression, and cognitive decline. The mind and body begin to spiral together in a cycle of mutual distress—mental stress feeding physical inflammation, and inflammation feeding emotional instability.

However, this cycle can be reversed. The first step is awareness: recognizing that stress is not only emotional but biological. Every anxious thought, every unresolved worry, sends ripples through the immune system. When we learn to calm the mind, we lower the inflammatory signals the brain sends to the body. This is why stress-reduction practices—meditation, slow breathing, yoga, nature walks, journaling—are not luxuries but biological necessities. They literally change your immune chemistry.

Scientific research supports this connection. Studies from Harvard, UCLA, and Johns Hopkins have shown that mindfulness meditation can lower levels of interleukin-6 (IL-6), one of the body's key inflammatory markers. Controlled breathing techniques activate the vagus nerve, stimulating the parasympathetic nervous system and shifting the body toward recovery. Gratitude and compassion practices have been linked to higher levels of immunoglobulin A (IgA), an antibody that serves as the immune system's first line of defense. These are not psychological effects—they are measurable physiological changes.

To retrain the stress response, consistency is more important than intensity. Just five to ten minutes of daily practice can begin to rewire neural pathways associated with calmness and resilience. Over time, the body learns to interpret life differently. A traffic jam no longer triggers a cortisol surge. A disagreement no longer feels like a physical threat. As your perception changes, your biochemistry changes.

Nutrition also plays a key role in managing inflammation. Diets high in processed foods, sugar, and trans fats fuel inflammatory pathways. On the other hand, foods rich in omega-3 fatty acids, antioxidants, and phytonutrients help calm overactive immune cells. Turmeric, green tea, berries, leafy greens, and olive oil are known for their anti-inflammatory properties. But even the healthiest diet can't compensate for an untrained mind. If your brain remains trapped in a cycle of fear or resentment, inflammation persists.

The goal, therefore, is not to eliminate stress but to transform your relationship with it. When you reframe stress as a signal for growth instead of danger, your brain interprets it differently, and your immune system follows suit. A stressful situation becomes an opportunity to build resilience instead of a threat to survival. This shift in mindset is not abstract—it has measurable effects on cortisol, heart rate, and immune function.

In essence, chronic inflammation is the body's way of expressing unresolved mental tension. It is your biology's way of saying, "Something within you needs to change." When we listen to that message and begin training the mind to respond rather than react, we move from survival to healing. The mind becomes the mediator between external chaos and internal peace, teaching the immune system to find balance once again.

Daniel's path to recovery

When Daniel's health began to decline, he blamed it on age. At forty-five, he was constantly tired, his joints ached, and his doctor told him his blood pressure was creeping upward. The diagnosis wasn't dramatic—just "chronic inflammation" and "stress-related fatigue." But Daniel couldn't understand how stress could make him physically sick. He was used to it. He ran a successful company, worked long hours, and always believed he could push through anything.

It wasn't until a minor cold turned into pneumonia that he started to pay attention. Lying in the hospital bed, connected to an IV, Daniel finally admitted to himself that something deeper was wrong. His mind never rested. Even when he was home, he was thinking about deadlines, clients, and competition. His body, it seemed, had finally had enough.

After being discharged, his doctor referred him to a wellness center that focused on stress-related disorders. During his first session, the therapist—a soft-spoken woman named Mariela—explained the science behind his symptoms. "Your immune system and brain have been in constant communication," she said. "Every thought of pressure or fear triggers a cascade of stress hormones. Over time, your body doesn't know when to stop fighting."

She drew a simple diagram showing how cortisol and adrenaline, meant to protect him, were actually breaking him down. "You've been living in survival mode for years," she added gently. "It's time to teach your body that it's safe again."

Daniel was skeptical. He'd built his career on discipline and productivity, not meditation or self-care. But he agreed to try. His first task was to spend ten minutes a day sitting quietly and focusing on his breathing. The first few sessions were torture. His thoughts wouldn't stop racing, and the silence made him restless. But Mariela told him to keep going. "You're not trying to stop thinking," she said. "You're training your nervous system to trust calmness again."

Weeks passed, and Daniel began to notice subtle shifts. He started sleeping through the night. The constant tightness in his chest lessened. He found himself pausing before reacting to stress. At work, he stopped rushing from one crisis to another and began delegating tasks he once micromanaged. It wasn't just his mind that changed—his body followed. His doctor noted lower inflammatory markers and improved heart rate variability, a sign of reduced stress.

One day, during a follow-up session, Mariela asked him to recall a moment of deep gratitude. Daniel thought of his daughter's laughter. He closed his eyes, pictured her face, and felt warmth spread through his chest. Mariela explained that this feeling—genuine gratitude—was one of the fastest ways to lower inflammation and boost immunity. "Your body listens when you feel safe," she said. "Gratitude tells your immune system the battle is over."

Months later, Daniel's transformation was visible. His posture straightened, his eyes looked clearer, and his energy returned. When colleagues asked what supplement he was taking, he smiled. "I just learned how to breathe again," he would say. But the truth was deeper: he had learned how to think differently.

He no longer viewed stress as his enemy but as a signal. Whenever he felt tension rising, he would pause, breathe, and remind himself that he had control over his response. The inflammation that once ruled his body had subsided because the fear that fueled it had dissolved.

Daniel's story became a testament to a profound truth—that chronic disease often begins not in the body, but in the mind. By retraining his thoughts, he rewrote his biology. His healing was not sudden, but steady, built one calm breath at a time.

Now, when Daniel looks back at his old life, he sees it clearly: the late nights, the endless worry, the self-imposed pressure. All of it had been inflammation in human form. Today, his immune system is no longer fighting him. It's working with him, guided by a brain that finally understands that peace is the strongest medicine of all.

Chapter 3: Neuroinflammation: When the Brain Becomes Inflamed

When people think of inflammation, they imagine swollen joints or an infection site—something visible and external. But one of the most critical and least understood forms of inflammation happens within the brain itself. Neuroinflammation is the silent disruptor behind many chronic conditions, including depression, anxiety, Alzheimer's disease, and chronic fatigue syndrome. It is the hidden bridge between emotional stress and physical illness—a state where the brain's own immune cells turn against it, creating a biochemical storm that affects every thought, emotion, and bodily function.

Inside the brain are specialized immune cells called microglia. These cells act like sentinels, constantly patrolling neural pathways for signs of infection, toxins, or cellular damage. When a threat is detected, they activate and release inflammatory molecules to neutralize it. This is a healthy and necessary process—until it doesn't turn off. Chronic psychological stress, poor diet, sleep deprivation, and environmental toxins can keep microglia in a constant state of activation. Over time, this low-grade neuroinflammation damages neurons, disrupts communication between brain regions, and alters mood regulation.

Cytokines—the chemical messengers of inflammation—play a key role here. When produced in excess, they interfere with neurotransmitters like serotonin, dopamine, and glutamate. This is why people with neuroinflammation often experience not only brain fog and fatigue but also emotional changes: irritability, depression, anxiety, and loss of motivation. The mind becomes clouded, not metaphorically but biologically. The brain's chemical balance is literally skewed by inflammatory signaling.

Another major driver of neuroinflammation is the gut-brain axis. The gut and brain are directly connected through the vagus nerve and communicate constantly. An imbalanced gut microbiome—often caused by processed foods, sugar, antibiotics, or chronic stress—leads to increased intestinal permeability, known as "leaky gut." This allows toxins and bacteria to enter the bloodstream, triggering immune responses that reach the brain. The result is a feedback loop: inflammation in the gut fuels inflammation in the brain, and vice versa. Restoring gut health is therefore a crucial step in protecting mental clarity and emotional stability.

Sleep also plays an essential role in preventing neuroinflammation. During deep sleep, the brain's glymphatic system activates—a cleansing process that removes waste products and inflammatory molecules. Chronic sleep deprivation shuts down this detoxification mechanism, allowing toxic proteins like beta-amyloid to accumulate. This buildup contributes to neurodegenerative conditions and impairs the brain's ability to regulate mood and memory. Thus, sleep is not simply rest—it is neurological repair.

One of the most powerful ways to reduce neuroinflammation is through consistent mental and physical training. Mindfulness, gratitude, and controlled breathing activate the parasympathetic nervous system and signal the brain to release anti-inflammatory compounds. Exercise increases blood flow and releases brain-derived neurotrophic factor (BDNF), which helps neurons grow and repair themselves. A plant-based, anti-inflammatory diet rich in omega-3s, polyphenols, and antioxidants strengthens this effect further.

Equally important is emotional regulation. Unresolved anger, resentment, or fear can keep microglia activated for years. When we replay old memories or harbor ongoing emotional conflict, the brain reacts as though the event is happening in real time. Emotional healing—through therapy, forgiveness, and self-awareness—literally quiets the immune system in the brain.

Reducing neuroinflammation is not about erasing stress but about improving communication within the body. When the mind and immune system begin to cooperate, microglia return to their resting state, neurons regenerate, and the fog lifts. The result is sharper cognition, deeper calm, and a renewed sense of vitality. By caring for your brain's immune environment, you are strengthening the foundation of every other system in your body. A peaceful brain creates a peaceful body—and a peaceful body creates the chemistry of long-term health.

How Maria learned to care for her mind

For years, Maria had lived with a cloud over her mind. It wasn't depression exactly, nor was it exhaustion alone. It was something deeper—a dull mental haze that seemed to follow her everywhere. She forgot names, misplaced her keys, and sometimes stared at her computer screen unable to recall what she was doing. Her doctor called it "stress-related fatigue" and recommended rest. But rest didn't help.

Maria worked as a graphic designer and prided herself on creativity, yet lately, even simple projects felt impossible. She drank more coffee, slept less, and pushed harder, convinced she could think her way out of the fog. The harder she tried, the worse it became. Headaches, irritability, and memory lapses crept in. When she snapped at her husband for forgetting groceries, she saw the fear in his eyes and realized something inside her had changed.

A close friend suggested she see a functional neurologist—a specialist who understood how lifestyle and emotion affect brain chemistry. During their consultation, the doctor explained that her symptoms were consistent with neuroinflammation. "Your brain's immune cells are overactive," he said. "They're supposed to protect you, but right now they're in attack mode. It's like there's a fire in your brain that never goes out."

He showed her a diagram of microglia and cytokines and how stress, diet, and lack of sleep could keep them inflamed. "When you stay in survival mode," he explained, "the brain interprets everything as a threat. Over time, that inflammation affects memory, focus, and mood. You're not broken—your brain is just tired of being on defense."

Maria began to cry. For years she had blamed herself for being unmotivated or weak, but now she understood the biology behind her struggle. The doctor gave her a plan: eliminate processed foods, take omega-3 supplements, get at least seven hours of sleep, walk daily, and practice fifteen minutes of deep breathing twice a day. Most importantly, he told her to slow down and learn how to rest her mind.

The first week was difficult. Her mind resisted stillness. When she tried to meditate, her thoughts darted everywhere. But she persisted. Slowly, the fog began to lift. She started noticing small victories: remembering a client's name, finishing a project on time, laughing more easily. Her husband noticed the change before she did. "You seem lighter," he said one evening. "It's like your eyes are brighter."

After two months, her follow-up tests showed reduced inflammatory markers. The doctor explained that her brain was finally recovering. "Your microglia are calming down," he said with a smile. "They're learning to trust you again."

Maria didn't realize how much she had changed until one afternoon when she was walking through a park, noticing how vivid the colors of the trees seemed. For years, she had been living through a dull filter, unaware of how much inflammation had dulled her senses and emotions. Now, everything looked alive again.

She still faced stressful days, but she approached them differently. When tension rose, she closed her eyes and took slow, deliberate breaths until she felt her body settle. She replaced late-night work sessions with evening walks and turned her morning coffee into green tea. Bit by bit, she rebuilt her mental clarity and emotional balance.

Years later, Maria became an advocate for brain health. She began speaking at wellness retreats about neuroinflammation and the power of mental training. “Healing my brain was like cleaning fogged glass,” she would tell audiences. “I could finally see myself again.” Her story reminded others that what we call “stress” or “mood swings” is often a biological imbalance—a brain inflamed by years of neglect and overwork.

Today, Maria no longer fears losing her focus. She knows how to care for her mind the way others care for their bodies—with patience, nourishment, and compassion. In healing her brain, she discovered something profound: clarity is not just a state of mind, it is a state of cellular harmony. And that harmony begins when we choose to make peace with ourselves.

Chapter 4: Emotional Immunity: How Feelings Shape Physical Health

Every emotion you experience creates a physiological ripple throughout your body. Joy, anger, fear, and love are not just psychological events—they are chemical and electrical states that directly shape your immune system's behavior. Emotions alter heart rate, hormone production, blood flow, and even gene expression. In this sense, emotional mastery is not simply a mental discipline—it is immune training at the deepest level.

When you feel stress or anger, your body releases cortisol and adrenaline, preparing for survival. These hormones, if sustained, suppress immune function and promote inflammation. When you experience gratitude or compassion, however, your body produces oxytocin, serotonin, and endorphins—chemicals that strengthen immunity, reduce inflammation, and promote healing. Each feeling you cultivate becomes a biochemical instruction to your cells, guiding them toward either defense or repair.

The concept of "emotional immunity" refers to the body's ability to remain balanced and resilient under emotional stress. It is not about suppressing emotions, but about regulating them intelligently. Emotional suppression—pretending to be calm while internally seething—has been shown to increase inflammation and elevate blood pressure. Emotional regulation—acknowledging feelings without letting them control your behavior—creates coherence between the brain and the heart, a physiological state linked with improved immune response and hormonal balance.

At the heart of this relationship lies the vagus nerve—the communication highway between the brain, heart, and immune system. When the vagus nerve is stimulated through calm breathing, laughter, singing, or meditation, it sends a signal of safety throughout the body. This activates the parasympathetic nervous system, lowering cortisol and reducing inflammatory cytokines. Over time, strengthening vagal tone becomes one of the most effective ways to maintain immune resilience and emotional stability.

Research has repeatedly confirmed this link between emotion and immunity. In one study, participants who practiced daily gratitude journaling for eight weeks showed a 23% decrease in stress hormones and higher levels of immunoglobulin A (IgA), an antibody essential for fighting infections. Other studies found that people who cultivate optimism and social connection recover faster from illness, respond better to vaccines, and experience lower levels of chronic inflammation.

Positive emotions literally reprogram immune cell behavior at the genetic level, improving their responsiveness and lifespan.

However, emotional immunity is not built through denial or forced positivity. True emotional strength comes from integration—the ability to experience emotions fully, extract their message, and release them. Every emotion carries information. Fear signals the need for caution. Anger signals that a boundary has been crossed. Sadness invites reflection and renewal. When we honor emotions instead of rejecting them, we prevent them from turning into chronic physiological stress.

A practical way to build emotional immunity is through mindfulness and reflection. When a strong emotion arises, pause before reacting. Observe the sensation in your body: the tightening chest, the quickening pulse, the heat in your face. Instead of resisting, breathe into it. This conscious acknowledgment allows your nervous system to process the emotion safely. Over time, your body learns that emotional waves are not threats—they are energy in motion.

Forgiveness is another powerful immune regulator. Resentment and bitterness sustain stress chemistry, keeping the body trapped in a state of vigilance. Forgiveness, on the other hand, releases that biochemical burden. Studies show that people who practice forgiveness have lower blood pressure, stronger immune markers, and reduced levels of inflammatory proteins. Letting go is not weakness—it is liberation for your biology.

Developing emotional immunity means creating a body that feels safe even when life feels uncertain. It means transforming emotional storms into signals for self-regulation and growth. The immune system thrives in an internal environment of peace, trust, and connection. By learning to manage emotions consciously, you send your cells a simple yet powerful message: *we are safe, we can heal, and we are whole.*

Asha's Feelings

Aisha was known for her composure. At work, she was calm under pressure, always the one her team turned to during crises. But behind her composed exterior, she carried an invisible weight. Years of holding in frustration—arguments left unresolved, disappointments swallowed silently—had begun to take their toll. Her immune system started to falter. She caught colds easily, her energy waned, and unexplained rashes appeared on her skin. Doctors ran tests, but everything looked "normal."

One day, while waiting for yet another specialist appointment, Aisha stumbled upon a lecture on "The Biology of Emotion."

The speaker, a neuroscientist, explained that chronic emotional suppression leads to inflammation. “Unexpressed emotions,” he said, “don’t disappear. They stay in the body, manifesting as fatigue, tension, and illness.” The words struck her like lightning. She realized that her body was expressing everything her voice never had.

That evening, Aisha began journaling. She wrote without censorship, pouring out years of resentment, anger, sadness, and fear. For the first time, she saw how much she had internalized. She wasn’t weak or broken—she was overloaded. As she wrote, she noticed her breathing deepen, her shoulders drop, and a subtle sense of relief settle over her body.

The next week, she joined a guided meditation group that focused on emotional release and self-compassion. During the sessions, the instructor taught a breathing exercise to activate the vagus nerve: a slow inhale through the nose, a gentle exhale through pursed lips, followed by a pause. “This tells your body you are safe,” the instructor explained. “When your body feels safe, your immune system begins to repair.”

At first, the idea seemed abstract. But after several weeks, Aisha noticed her chronic tension easing. Her digestion improved, her skin cleared, and her sleep deepened. She began smiling more often—genuine, unforced smiles that came from within. She learned that calmness wasn't the absence of emotion—it was the integration of it.

A turning point came when she revisited a painful memory from years before—a betrayal by a close friend. For years, she had held onto bitterness, replaying the scene in her mind whenever she felt hurt. This time, guided by her meditation coach, she chose to visualize the moment differently. She saw herself releasing the anger, not because the betrayal was acceptable, but because holding it no longer served her. Tears streamed down her face as she whispered, "I forgive you." In that moment, something inside her shifted.

In the days that followed, Aisha felt lighter, as if her body had shed years of invisible armor. Her next blood test revealed a surprising improvement—her inflammatory markers had dropped significantly. Her doctor, puzzled, asked what she had done differently. "I just stopped fighting my feelings," she said softly.

Months later, Aisha began leading workshops for others on emotional healing and immunity. She shared her story openly, explaining that emotions are not enemies to be suppressed, but messengers to be understood. "When you honor what you feel," she told her students, "your body stops screaming for your attention."

Her life transformed—not because she eliminated stress, but because she learned to meet it with awareness instead of resistance. She still faced challenges at work, family tensions, and moments of sadness, but her body no longer reacted with inflammation. Instead, she breathed through them, knowing each emotion was simply energy passing through her system.

Today, Aisha's calmness radiates differently. It's not the rigid control she once used to mask her pain, but a relaxed confidence rooted in self-acceptance. She understands that true strength is not about never breaking—it's about knowing how to mend. Her health, once fragile, became a reflection of her inner peace.

Through her journey, Aisha discovered what science has only recently begun to prove: emotions are not abstract—they are molecular. By transforming her emotional life, she rewrote her biology. Her immune system didn't just heal; it evolved. And with every deep breath and act of forgiveness, her body continued to whisper the same message her heart had always known: *peace is the strongest immunity of all.*

Chapter 5: The Mind–Body Reset: Rewiring Health Through Conscious Awareness

The body's default state is health. Every cell is designed to maintain equilibrium—to repair, regenerate, and adapt. Yet most people live in a state of chronic imbalance because the mind constantly sends mixed or harmful signals to the body. This state of disconnection is the true root of many chronic illnesses. The key to reversing it lies in learning how to consciously reset the mind–body connection so that both systems work together rather than in opposition.

Your brain is not only the control center of your nervous system—it is also the translator of your inner dialogue. Every thought you have becomes a biochemical signal that tells the body how to respond. When you think fearful or self-critical thoughts, the brain interprets danger and triggers stress chemistry. When you think thoughts of safety, gratitude, or trust, the brain activates healing chemistry. This process occurs continuously, shaping your immune function, digestion, cardiovascular health, and hormonal balance.

The "mind–body reset" begins with awareness. You cannot change what you do not notice. Many people live in automatic patterns of tension, anxiety, or overthinking, unaware that their mental state is instructing their body to remain in survival mode. The first step to healing is to observe these unconscious loops. Notice how your shoulders tighten when you feel pressure, or how your breathing becomes shallow when you anticipate conflict. These are signs that your nervous system is misinterpreting daily life as danger.

Once awareness is established, the next step is regulation. Conscious breathing is the most direct way to reset the body's physiology. A simple exercise known as the 4–6 breathing technique—inhaling for four seconds and exhaling for six—stimulates the vagus nerve and activates the parasympathetic nervous system. Within minutes, heart rate slows, blood pressure drops, and cortisol decreases. Practicing this several times per day re-trains the brain to associate calmness with safety, gradually reprogramming the body's stress response.

Movement and posture are also essential to the mind–body reset. When the body is slumped or tense, it reinforces emotional heaviness and fatigue. Standing tall, stretching the spine, and practicing slow, rhythmic movement—like yoga, tai chi, or mindful walking—send signals of confidence and safety to the brain. The nervous system mirrors the body's position.

Changing posture changes perception, which in turn changes biochemistry.

Visualization and intentional thought are the next layers of reprogramming. The brain cannot distinguish vividly imagined experiences from real ones, which means visualization can be used to direct healing. Studies have shown that guided imagery can reduce pain, accelerate wound healing, and improve immune responses. When you visualize your organs functioning perfectly, your immune cells defending effectively, or your body filled with light and balance, you are sending a biological instruction that aligns neural and immune activity toward restoration.

To reinforce this transformation, it is important to release stored emotional energy. Emotional memory lives in the body as muscle tension, shallow breathing, or chronic pain. Techniques such as journaling, breathwork, and somatic release exercises help free the body from these old imprints. This not only relieves physical discomfort but also clears mental fog, creating space for new patterns of calm and vitality.

The final step in the mind–body reset is consistency. Change occurs through repetition. Just as the brain can be trained into anxiety and disease through years of stress, it can be retrained into peace and resilience through daily practice. Begin each morning with stillness—breathe, center yourself, and set the intention for calm awareness. Throughout the day, pause often, check in with your body, and correct your posture or breathing whenever tension arises. Over time, your nervous system will learn to stay balanced automatically.

The mind–body reset is not about control—it's about communication. When you learn to listen to your body's signals and respond with awareness, you enter a partnership with your biology. Healing then becomes a conversation, not a struggle. By consciously sending signals of safety, peace, and trust, you restore your body's natural capacity to heal itself. Health is not achieved by force—it is remembered through awareness.

Rewiring Ethan

Ethan had always considered himself healthy. He worked out regularly, ate well, and prided himself on discipline. But over time, his body began to rebel. Sleep became restless, digestion unpredictable, and his once-steady energy disappeared. His doctor prescribed supplements and suggested reducing caffeine, but nothing changed. Deep down, Ethan suspected the problem wasn't physical—it was mental. He lived with constant tension, even when there was nothing wrong. His mind was always running, his body always braced.

One evening, after another sleepless night, he came across a podcast episode called *"Rewiring the Mind–Body Connection."* The guest, a neuroscientist, explained how chronic stress creates a feedback loop between the brain and the body, locking people into a state of survival. "You can't heal in the same state of mind that made you sick," she said. "You have to teach your body what safety feels like again."

That sentence echoed in his head for days. Teach the body safety. He realized he had been living as though danger were everywhere—racing thoughts, tight deadlines, rushing meals, pushing workouts beyond exhaustion. His body wasn't malfunctioning; it was protecting him from his own pace of life.

Determined to change, Ethan signed up for a mindfulness and breathwork workshop. On the first day, the instructor asked participants to close their eyes, breathe slowly, and simply feel their bodies. Within seconds, Ethan felt a wall of discomfort. His mind wanted to move, fix, analyze. The stillness was unbearable. The instructor's voice was calm: "Don't fight it. Just observe. Awareness itself begins the reset."

Over the following weeks, Ethan practiced daily. At first, it felt mechanical—inhale for four seconds, exhale for six. But soon, something subtle began to shift. The mornings that once started with tension began with calm. His thoughts slowed. His body, once tight, started to soften. He noticed that when he stayed aware of his breath, his digestion improved and his energy lasted longer.

Encouraged, he began experimenting with visualization. Each night before bed, he imagined his brain and body communicating clearly—his mind sending signals of peace, his immune system responding with harmony. He pictured light traveling through his body, clearing tension from his chest and stomach. To his surprise, he began sleeping deeply for the first time in years.

One day, while hiking, he caught himself walking slowly, appreciating the rhythm of his steps and the sound of wind through the trees. For the first time in memory, he wasn't racing toward a goal. He was simply present. His body felt lighter, as if it had been waiting years for him to stop fighting it.

After several months, Ethan's transformation was undeniable. His blood pressure dropped, his digestion normalized, and the chronic fatigue vanished. But the biggest change wasn't measurable—it was how he felt inside. The anxiety that once lived in his chest had turned into a quiet confidence. "My body finally trusts me again," he told a friend.

Curious, Ethan began studying more about the science behind his experience. He learned that his consistent breathing and mindfulness practices had increased his vagal tone, rebalanced his HPA axis, and lowered inflammatory markers. His nervous system had literally reprogrammed itself to interpret life as safe rather than dangerous.

Later, Ethan became a wellness coach, teaching others how to reconnect their minds and bodies. He shared his story openly: "I used to think health was about control—pushing harder, eating cleaner, working longer. But real health began the moment I stopped controlling and started listening."

He taught his clients the same breathing exercise that changed his life. Some were skeptical at first, as he had been. But after weeks of practice, they too reported better sleep, fewer headaches, and calmer moods. Ethan knew then that healing wasn't something you achieved—it was something you allowed.

Now, every morning, he sits quietly before starting his day, hands on his chest, eyes closed. He breathes deeply, feeling his heartbeat sync with his breath. "I am safe," he whispers. "I am at peace." His body responds instantly, as if whispering back, "I know."

Through his journey, Ethan learned that the mind and body are not separate entities but reflections of each other. When one heals, the other follows. His story became proof that awareness is not passive—it is the most powerful medicine of all. By teaching his body the language of calm, he unlocked the natural intelligence within him that had been waiting all along to restore balance, energy, and life.

Chapter 6: The Healing State: Activating the Body's Natural Repair Systems

Healing is not something you force—it is something you allow. The human body is designed to heal itself. Every day, cells repair DNA damage, immune cells destroy pathogens, and tissues regenerate. Yet when the mind and body are trapped in a chronic state of stress or fear, the healing mechanisms shut down. The energy that should go toward repair is diverted to defense. The "healing state" is the biological opposite of the "stress state." It is the mode in which the body restores itself, guided by signals of safety, trust, and peace.

When the body perceives safety, the parasympathetic nervous system takes control. Heart rate slows, blood pressure drops, digestion improves, and inflammation subsides. The immune system shifts from attack mode to maintenance mode, repairing damaged tissues and restoring balance. Hormones like serotonin and oxytocin are released, promoting calm and emotional connection. The mind becomes clear, and the body begins to regenerate. This is the healing state—a physiological environment of restoration and balance.

One of the most direct ways to activate this state is through conscious relaxation. Most people think of relaxation as sitting still or watching television, but true relaxation is an intentional reset of the nervous system. Techniques like diaphragmatic breathing, guided meditation, or progressive muscle relaxation send a message to the body that it is safe to rest. Within minutes, stress hormone levels begin to fall. Repeated daily, this practice reconditions the body to enter the healing state naturally, even in stressful environments.

Visualization amplifies this process. Studies in psychoneuroimmunology have shown that when people visualize their bodies healing—imagine light moving through their organs, or cells regenerating—immune markers improve. The brain responds to imagery as if it were reality. When you vividly imagine yourself in a calm, healthy, and vibrant state, the body mirrors that image through biochemical responses. The mind paints the blueprint; the body builds according to it.

Another powerful pathway into the healing state is gratitude. Gratitude shifts brain activity from the amygdala—the center of fear—to the prefrontal cortex, which governs calm thinking and emotional regulation. This change in brain activity reduces cortisol and increases dopamine and serotonin. Practicing gratitude daily, whether through journaling or silent reflection, teaches the body to associate calmness with abundance rather than threat. Over time, gratitude becomes a biological signal that activates healing.

Sleep is another cornerstone of the healing state. During deep sleep, the body performs vital repair work: immune cells clear toxins, the liver processes waste, and the brain consolidates memory and cleans out inflammation through the glymphatic system. Without adequate rest, the body remains chemically imbalanced, unable to fully repair. Therefore, creating an evening routine—dim lights, slow breathing, and no screens—prepares the nervous system for recovery.

Nutrition and movement also play crucial roles. Foods rich in antioxidants, minerals, and healthy fats reduce oxidative stress and provide the raw materials for tissue repair. Gentle forms of exercise like stretching, walking, or tai chi increase circulation and oxygenation, both essential for healing. But equally important is intention—approaching movement not as punishment or effort, but as gratitude for the body's ability to move.

The final and perhaps most profound key to entering the healing state is surrender. Healing cannot occur in the presence of resistance. The moment you stop fighting your body and begin trusting its intelligence, energy returns to where it belongs—toward repair. The act of surrender is not giving up; it is giving permission. When the mind releases the need to control, the body remembers how to heal.

Healing is not a mystery—it is the body's natural rhythm. The goal is not to create healing, but to remove what blocks it: fear, tension, self-criticism, and disbelief. When you replace those with calm, awareness, and trust, your biology follows. In the healing state, time slows down, breath deepens, and the body's wisdom awakens. This is where true regeneration begins—not in medication or struggle, but in the stillness where the mind finally says to the body, "It's safe now—you can begin."

How Sofia found her ideal guided healing state

For years, Sofia lived in constant motion. She ran a successful business, raised two children, and balanced a dozen responsibilities at once. But beneath the surface, her body was breaking down. Persistent fatigue, stomach pain, and insomnia had become her new normal. She tried supplements, diets, and new routines, but nothing worked. Deep inside, she felt her body was pleading for something she didn't understand.

One afternoon, after collapsing from exhaustion during a meeting, Sofia's doctor told her she had adrenal fatigue and early autoimmune markers. "You need to slow down," he said firmly. "Your body's healing systems can't function while you're in survival mode." The words hit her harder than any diagnosis. She wasn't just tired—she was living in constant fight-or-flight.

Determined to recover, she began researching the connection between stress and healing. She learned that her nervous system had been stuck in high alert for years. Every morning she woke up to her phone's alarms, her heart already racing, her mind listing tasks. Her body never received the signal that it was safe to rest.

One evening, while scrolling online, she came across a guided audio session titled *"Activate Your Healing State."* The voice on the recording spoke slowly: "Your body knows how to heal. All it needs is your permission." The words brought tears to her eyes. She followed the instructions—closed her eyes, took a deep breath, and imagined a warm golden light expanding from her chest through her entire body. For the first time in years, her breathing slowed naturally. When the session ended, she felt something she hadn't felt in a long time—peace.

She began practicing this every morning and night. At first, it seemed too simple to make a difference. But after a week, she noticed she was sleeping through the night. The pain in her stomach began to fade. She smiled more easily and felt her energy returning. Her doctor was astonished by her next lab results—her inflammation markers had dropped, and her immune function had improved significantly.

Encouraged, Sofia started integrating other healing rituals into her day. She took short walks outside, wrote down three things she was grateful for each night, and practiced slow breathing whenever she felt overwhelmed. Over time, she realized these weren't just habits—they were her body's way of communicating safety to her mind. The more she trusted the process, the faster she healed.

Months later, while sitting in her garden one morning, she had a moment of deep clarity. Watching the sunlight filter through the trees, she realized that healing was not about doing—it was about being. Her body hadn't needed more effort; it had needed her presence. Every breath, every pause, every quiet moment was a signal to her cells that life was safe again.

Her transformation inspired those around her. Friends who once saw her as the unstoppable entrepreneur now admired her serenity. "You're different," one friend said. "You seem... healed." Sofia smiled. "I stopped fighting myself," she replied. "And once I did, my body did the rest."

In the years that followed, Sofia began teaching workshops on "The Healing State." She guided others through the same process—helping them slow down, breathe, and reconnect. She told them what she had learned: "Your body is not against you.

It's always trying to protect you. But you have to show it that the danger has passed."

Her story became a living example of the power of surrender. She had gone from exhaustion to vitality not by adding more tasks, but by removing the inner resistance that kept her body trapped in survival. Healing, she realized, wasn't an event—it was a relationship. A conversation between her thoughts and her cells, her breath and her heartbeat.

Today, Sofia wakes before sunrise, sits quietly with her eyes closed, and breathes slowly until she feels her heartbeat soften. "I trust you," she whispers to her body. In that silence, she feels the same golden warmth she first discovered months ago—a reminder that healing is not found in effort, but in stillness.

Her journey taught her that the body's intelligence is vast, its capacity for repair limitless. When the mind stops sending signals of fear and begins radiating calm, every cell responds. The healing state is not something to find; it's something to return to—a state that has always been within us, waiting patiently for permission to begin.

Chapter 7: The Placebo Power: How Belief Triggers Biological Healing

Belief is not just psychological—it is biochemical. When you believe that healing is possible, your brain and body act as though it is already happening. This is the essence of the placebo effect: a powerful demonstration of how the mind can influence biology. For decades, scientists have observed patients recovering from illnesses or feeling profound relief after taking pills that contained no active ingredients. The only real medicine was belief itself. What was once dismissed as "just placebo" is now recognized as one of the most powerful healing forces known to science.

The mechanism behind the placebo effect lies in the brain's expectation system. When you expect healing, your brain releases chemicals that create the same physiological effects as real medication. Expectation becomes biology. Dopamine, serotonin, and endorphins surge; stress hormones drop; immune function improves. Brain scans have shown that placebos activate the same neural pathways as painkillers, antidepressants, and even heart medications. The mind's belief sends a clear message to the body: healing is underway.

The placebo effect is not about deception—it's about activating the body's internal pharmacy. Every person has within them the capacity to release chemicals that reduce pain, regulate inflammation, and accelerate recovery. The challenge is not accessing this pharmacy—it's believing in it. When we dismiss our own capacity for healing as "unscientific," we cut off communication between the brain and the body's innate intelligence. But when we combine faith, visualization, and positive expectation with medical treatment, we multiply the body's responsiveness to healing.

The opposite of the placebo is the *nocebo* effect—the negative expectation that something will harm us. Fear, doubt, and hopelessness produce biological changes just as real as hope does. People who believe they will get worse often do, even when their physical condition should improve. This demonstrates that the mind does not differentiate between belief in wellness or belief in illness—it simply obeys. Whatever image or emotion dominates your consciousness becomes your body's blueprint.

The key to using belief for healing lies in creating coherence between thought, emotion, and physiology. Merely repeating affirmations without feeling their truth does little. But when belief is felt emotionally—when gratitude, trust, and joy accompany the thought—your nervous system enters the healing state. The limbic system, which governs emotion, communicates directly with the hypothalamus and the immune system. A joyful belief in recovery is not wishful thinking; it's a neurochemical command that mobilizes the body's self-healing resources.

To activate this process, begin with visualization and sensory detail. Imagine your body healing in real time—tissues regenerating, energy flowing, inflammation fading. Feel the emotions associated with complete health. Gratitude is particularly potent because it anchors the belief that healing is already occurring. The brain interprets gratitude as confirmation that safety and abundance are present, releasing serotonin and oxytocin. This combination strengthens the immune system and lowers cortisol, allowing the healing process to accelerate.

The placebo power also reveals why hope and connection are vital in medicine. A caring doctor, encouraging words, and compassionate environments enhance patient outcomes. It's not only the treatment that heals—it's the emotional environment surrounding it. When someone feels seen, supported, and believed in, their body produces more nitric oxide, improving blood flow and cellular repair. The science of healing, therefore, is also the science of belief, trust, and human connection.

The true lesson of the placebo effect is empowerment. It

reminds us that healing is not something that happens *to* us—it's something that happens *through* us. By aligning mind, emotion, and biology, we awaken the body's internal healing systems. Every thought of hope is a molecular event. Every belief in recovery sends an electrical signal that reprograms your cells. You are not separate from your healing—you are its source.

Marcus, the believer

When Marcus was diagnosed with a chronic autoimmune disorder, his doctors told him it was "manageable but incurable." The medications helped with the pain, but the fatigue and inflammation persisted. Over time, Marcus's hope faded. He began to live cautiously—avoiding certain foods, avoiding exercise, and avoiding dreams of full recovery. He didn't realize that the greatest damage wasn't being done by his illness, but by his own belief that healing was impossible.

One day, a friend sent him a link to a documentary about the placebo effect. Skeptical but curious, Marcus watched as patients with similar diagnoses experienced dramatic improvement after participating in "mind-body healing experiments." The patients had received saline injections or sugar pills, but their pain vanished, their inflammation markers improved, and many regained full function. The researchers explained that belief had activated the brain's own pharmacy of healing chemicals. "Your body listens to your thoughts," one doctor said. That phrase stayed with Marcus long after the film ended.

For the first time in years, he wondered—what if he could train his mind to believe again?

Marcus began each morning by closing his eyes and imagining his body as vibrant, strong, and whole. He pictured his immune system recalibrating, his blood glowing with energy, and his cells communicating in harmony. At first, it felt ridiculous. He didn't *believe* it; he was just pretending. But he persisted. Within a few days, he began to notice subtle changes. His pain seemed less sharp, and he woke up feeling slightly lighter.

Encouraged, he started keeping a gratitude journal, listing five things each day that he appreciated. Over time, gratitude became a habit. He stopped focusing on what was wrong and began celebrating what was right—the way his body still moved, the sunlight in the mornings, the simple joy of breathing without pain. Each moment of gratitude deepened his trust in the body's intelligence.

During a routine checkup three months later, Marcus's doctor was surprised. His inflammation markers had dropped significantly, and his energy levels were improving. "Whatever you're doing, keep doing it," she said, puzzled. He smiled and simply replied, "I started believing again."

The more he believed, the more his body responded. He began visualizing his medications as catalysts for healing instead of chemical crutches. "This pill is just a messenger," he told himself. "The real healing is coming from within." The results were astounding. His symptoms continued to fade until he no longer needed high doses of medication.

One evening, while walking through a quiet park, Marcus had a realization that brought him to tears. All those years, he had been waiting for his body to prove it could heal before believing in it. But the truth was the opposite—his body had been waiting for him to believe first.

From that day forward, Marcus shared his story with others who felt hopeless. He taught small workshops about the power of expectation, helping people harness their minds as partners in healing. "You don't need blind faith," he would tell them. "You need conscious trust—the kind that your body can feel."

One of his most powerful experiences came when he met a young woman newly diagnosed with the same condition he once had. She was terrified and defeated. Marcus looked her in the eyes and said softly, "The same intelligence that keeps your heart beating knows how to heal you. Believe in that, even when it feels impossible." She later told him those words changed her life.

Years later, Marcus's own health remained stable. He wasn't just free of symptoms—he was stronger than before. His journey taught him that belief isn't about pretending everything is fine. It's about creating a space inside where healing can begin. His mind had once been his greatest obstacle; now it was his greatest ally.

Every morning, as he sits in silence, he repeats a simple affirmation: "My body knows what to do. I believe in its wisdom." He doesn't say it to convince himself anymore—he says it because he's living proof.

Marcus's story is not a miracle; it is biology aligned with belief. His recovery stands as a testament to the truth that the mind and body are partners, not separate entities. The placebo power is not about trickery—it is the natural intelligence of the body responding to faith. His journey revealed that healing begins not with medicine, but with meaning—and that the most potent prescription you can ever take is belief itself.

Chapter 8: Energy and Frequency: The Invisible Language of Healing

Every atom in your body vibrates. Every cell emits an electrical signal. Every organ resonates at a specific frequency that communicates with the rest of the body. Science has confirmed that humans are not just biological systems—we are also electromagnetic systems. The heart, brain, and nervous system constantly generate measurable electrical and magnetic fields that affect our physiology and the environment around us. Healing, therefore, is not only biochemical—it is energetic. The state of your inner frequency determines how efficiently your body repairs, regenerates, and communicates.

The heart's electromagnetic field is the most powerful in the body—about sixty times stronger electrically and up to five thousand times stronger magnetically than the brain's field. This means the heart is not just a pump; it is an information transmitter. Emotions such as love, gratitude, and compassion create coherent heart rhythms—smooth, harmonious waves that promote optimal communication between the brain and body. Negative emotions such as anger, fear, and frustration create chaotic heart rhythms, which disrupt this communication and trigger stress chemistry. Coherence between the heart and brain is the energetic foundation of healing.

When your heart rhythm becomes coherent, your entire body synchronizes. The nervous, hormonal, and immune systems align in balance, creating what researchers call "physiological resonance." This is the healing state at the frequency level. Achieving coherence can be as simple as focusing on your heart, breathing slowly, and cultivating a positive emotion such as gratitude. Within minutes, your brain waves shift into alpha or theta states—patterns associated with deep relaxation, creativity, and accelerated healing.

Energy medicine traditions around the world—Reiki, Qi Gong, acupuncture, sound therapy—have long worked with this principle. They recognize that when energy flow is blocked or imbalanced, illness manifests. Modern science now supports these ancient ideas through studies in biofield research, showing that changes in electromagnetic frequency correlate with pain, inflammation, and immune dysfunction. When balance is restored, the body naturally moves toward harmony.

Sound and vibration also play a major role in frequency-based healing. Every sound produces a vibration that interacts with your nervous system. Listening to soothing music, chanting, or using specific frequencies like 432 Hz or 528 Hz has been shown to reduce anxiety, lower heart rate, and increase parasympathetic activity. Similarly, spending time in nature exposes you to frequencies that help reset your body's electromagnetic rhythm—what researchers call "grounding." Walking barefoot on grass or soil allows the body to absorb the Earth's negative ions, which neutralize free radicals and reduce inflammation.

Energy healing, however, begins within. Your thoughts and emotions are frequencies. A positive mindset creates a coherent vibration that strengthens your biofield, while negative thinking creates distortion and energetic noise. This is why meditation and mindfulness are so effective—they clear the static and reestablish a clean signal between mind and body. Visualization amplifies this effect by adding direction to your energy. When you imagine light, warmth, or color flowing through your body, you are literally reprogramming your energetic system toward balance.

It is important to understand that energy healing is not mystical—it is measurable. Technologies like EEG, EKG, and heart-rate variability monitoring already track the body's electrical activity. These same signals shift dramatically when a person enters states of calm, gratitude, or love. The frequencies of healing are not abstract—they are real electromagnetic changes reflecting the body's inner harmony.

To live in alignment with healing frequency, make your daily environment resonate with calmness and life. Spend time in sunlight, listen to music that uplifts you, surround yourself with kind people, and nurture thoughts that empower you. Your energy field is constantly interacting with the world, amplifying what you focus on. When you cultivate a high-frequency inner state—gratitude, peace, compassion—your body follows suit. Healing becomes effortless because your energy, biology, and consciousness are finally speaking the same language.

Leila's migraines

For years, Leila struggled with chronic migraines that no doctor could fully explain. Tests came back normal, medications dulled the pain but left her drained, and specialists concluded it was "stress-related." Frustrated, she began exploring alternative methods of healing. That's when she discovered an online talk about energy frequency and heart coherence. The speaker, a physicist turned researcher, claimed that emotions have measurable vibrations—and that harmony within the heart and brain could restore balance to the entire body. Leila was skeptical, but her curiosity outweighed her doubt.

One evening, during a particularly severe migraine, she decided to try one of the exercises described in the talk. Sitting in a dark room, she placed her hand over her heart and began breathing slowly. The instruction was simple: inhale for five seconds, exhale for five seconds, and focus on a feeling of gratitude. At first, her mind wandered, replaying the day's frustrations. But after a few minutes, something shifted. Her breathing deepened, her muscles softened, and for the first time in months, the pressure in her head began to ease. She fell asleep not in pain, but in peace.

Over the next few weeks, Leila committed to this daily practice. Every morning, she sat quietly, breathed through her heart, and recalled something she loved—a sunrise, a memory of laughter, the sound of her daughter's voice. She didn't think of it as "therapy"; she thought of it as tuning her inner frequency. Gradually, the migraines became less frequent. When they did occur, she noticed they were less intense and often dissolved when she practiced coherence breathing.

Curious about the science, she purchased a small device that measured heart-rate variability—the same tool used in heart coherence research. Watching the screen, she could see her heart rhythm shift from jagged, erratic lines to smooth, wave-like patterns as she focused on gratitude. She realized she was witnessing her own frequency changing in real time. The connection between emotion and physiology was no longer abstract—it was visible.

Encouraged by her progress, Leila began experimenting with sound healing. She played soft frequencies—528 Hz and 432 Hz tones—while meditating. The gentle vibrations seemed to melt tension from her body. On weekends, she walked barefoot along the beach, feeling the pulse of the earth beneath her feet. "It's like my body is remembering something it forgot," she told a friend.

Months later, her migraines had nearly disappeared. Her doctor was astonished. "Whatever you're doing is working," he said. She smiled and replied, "I just learned how to listen to my body's frequency."

Leila's transformation went beyond physical healing. She noticed that her relationships improved, her mood stabilized, and her creativity flourished. The more she aligned her inner energy with peace and gratitude, the more her outer world reflected it. She began to see everything—including people—as energy in motion. Arguments, environments, even thoughts—all carried frequencies that could either strengthen or weaken her balance. She learned to protect her energy by choosing calm over chaos, kindness over anger, and silence over reaction.

One evening, while guiding a small group through heart-breathing meditation, Leila shared her journey. "I used to think healing was something that happened to the body," she said softly. "Now I know it's something that happens through energy. My thoughts, emotions, and breath are all part of one frequency. When they're aligned, my body follows."

As the group sat quietly, breathing in rhythm, Leila felt a wave of warmth radiate from her chest—a field of coherence she could almost sense in the room. In that stillness, she realized that healing was never about curing symptoms; it was about restoring harmony to the invisible music of the body.

Years later, she continued to live by this understanding. She taught that every word, every breath, every thought carries energy—and that the body is always listening. By choosing peace, gratitude, and compassion, she had retuned her entire being to the frequency of healing. Her migraines had been the body's way of asking her to listen to that deeper song. Now, she lived in rhythm with it—proof that healing is not just about what you do, but the vibration you live in every day.

Chapter 9: The Science of Surrender: Letting Go to Heal Faster

Healing often begins the moment we stop trying to control it. The modern mind is conditioned to fix, force, and fight, but the body heals through flow, not struggle. The science of surrender is the understanding that the body's intelligence far exceeds conscious thought—and that the more we relax our grip, the faster that intelligence can do its work. Surrender is not weakness; it is alignment. It is the act of trusting your body's natural rhythm instead of resisting it.

When the mind is in resistance, the body stays in tension. This tension blocks circulation, slows digestion, and keeps the nervous system locked in fight-or-flight mode. The result is delayed healing, chronic fatigue, and inflammation. From a physiological standpoint, control and fear activate the sympathetic nervous system—the survival branch—while surrender activates the parasympathetic branch—the repair branch. True recovery only happens when the body feels safe enough to let go.

Neuroscience confirms that letting go changes the brain. When we release attachment to outcomes and stop replaying worries, activity in the amygdala decreases, and the prefrontal cortex—the center for reason and peace—takes over. This shift reduces cortisol, lowers blood pressure, and restores balance to the immune system. The act of surrender is therefore not only spiritual but biological. It rebalances brain chemistry and reactivates the body's repair systems.

To practice surrender, begin with awareness of resistance. Notice where in your body you feel tightness—your shoulders, chest, or stomach—and ask yourself what thought or fear is creating that tension. Then take a slow, conscious breath into that space. On the exhale, silently say, *I release what I can't control.* This simple act signals the nervous system that it's safe to relax. Over time, this response becomes automatic.

Visualization can deepen surrender. Imagine a current of light or water flowing through your body, washing away all pressure and expectation. See your worries dissolving like mist as you exhale. The brain responds to imagery, and when it sees safety and flow, it creates the internal chemistry of healing. Studies on meditation practitioners show that visualizing surrender lowers inflammatory markers and strengthens the vagal tone—the body's signal for safety and resilience.

Surrender also has a profound effect on immune regulation. When we resist life—through fear, guilt, or obsession—we create internal conflict. The immune system mirrors that conflict, attacking or overreacting to perceived threats. When we let go, internal harmony returns, and the immune system recalibrates. This is why people who adopt acceptance-based mindsets often recover more quickly after surgery, trauma, or illness. Acceptance is not resignation—it's cooperation with the body's wisdom.

Faith and trust amplify this process. Whether you view healing through a spiritual, scientific, or personal lens, belief in the body's ability to restore itself changes biology. Studies have shown that patients who hold hope and faith recover faster and report less pain. The reason is simple: trust relaxes the nervous system, which enhances blood flow, immunity, and cellular repair. Surrender is the ultimate signal of faith—it tells the body, *I am no longer in danger; you may heal now.*

Letting go does not mean doing nothing. It means doing what you can from a place of calm rather than panic, then releasing the need to control the outcome. When the mind stops forcing, energy begins flowing. The body's inner physician awakens. Cells, hormones, and neurotransmitters respond as if the storm has passed. This is the science of surrender—the transformation of tension into trust, of control into flow, of resistance into recovery.

Healing accelerates when we finally allow it to happen. The body always knows how to return to balance—it simply waits for permission. Surrender gives that permission. When you release fear, the body releases pain. When you let go of control, the heart opens. And in that space of trust, the natural intelligence of life begins its quiet, miraculous work.

The moment Jenna decided to let go

For most of her life, Jenna believed that strength meant control. She planned every detail, organized every moment, and pushed through pain with determination. It served her well in her career but left her body exhausted. When she was diagnosed with a stress-induced heart condition at forty-three, her world collapsed. "You need to slow down," her doctor said gently. "Your body can't keep living in overdrive." But slowing down felt impossible. Her identity was built on doing, fixing, and achieving.

At first, Jenna tried to *manage* her recovery the same way she managed her work—strict routines, perfect diets, endless research. Yet the more she tried to control her healing, the worse she felt. Her heart palpitations persisted, and her anxiety grew. She couldn't understand why she wasn't improving. One night, unable to sleep, she stumbled upon a podcast about the "science of surrender." The speaker described how true healing happens when the body feels safe enough to let go. The phrase echoed in her mind: *safe enough to let go.*

For the first time, Jenna realized she had never truly felt safe—not with herself, not with life. Everything had been a battle for control. That night, she sat in silence, closed her eyes, and whispered, "I surrender." The words felt foreign, almost frightening. But as she breathed deeply, something inside her began to soften. She didn't feel better instantly, but she felt *different*. It was as if her body had been waiting for those words for years.

Over the next few weeks, Jenna made a conscious effort to stop forcing. When anxiety surged, she breathed through it instead of fighting it. When her thoughts raced, she whispered her mantra: "I release what I can't control." She began journaling, not to solve her problems but to express them and let them go. Each day, the tension in her chest lessened. Her sleep deepened. Her body, once rigid, began to feel light and fluid again.

One morning, while walking through a park, Jenna noticed something she hadn't in months—the sound of birds, the warmth of the sun on her face. Tears filled her eyes as she realized she was present for the first time in years. That evening, she practiced a guided meditation on surrender. The instructor said, "Healing is what happens when you stop trying to heal." Jenna felt a quiet truth resonate in her bones.

Her next doctor's visit revealed dramatic improvement. Her heart rate variability—an indicator of nervous system balance—had normalized. "Whatever you're doing," the doctor said, "keep doing it. Your body is healing itself." Jenna smiled. She didn't say much, but inside she knew the truth—she hadn't *done* anything. She had simply stopped resisting.

Months later, Jenna's life looked completely different. She still worked, but she no longer lived in panic or perfection. She approached challenges with calm curiosity instead of urgency. When setbacks arose, she reminded herself, "Flow, not force." Her friends noticed the change too. "You seem peaceful," one said. "Like you're not carrying the world anymore." Jenna laughed softly. "Because I finally realized I never had to."

In time, she began sharing her story with others who were struggling with stress and burnout. She told them that surrender wasn't giving up—it was giving back control to the part of the body that always knew what to do. "When you let go," she explained, "your body finally gets the message that the war is over. That's when healing begins."

One evening, Jenna sat on her porch, breathing slowly, watching the sky turn from gold to purple. The quiet hum of crickets surrounded her, and she felt her heartbeat steady and calm. For years, she had tried to conquer her body. Now, she had learned to listen to it. Each breath felt like a conversation with life itself.

She closed her eyes and smiled, feeling that same inner peace she once thought was unattainable. Healing had not come through effort or control—it had arrived the moment she trusted her body to remember what to do. In that surrender, she found something she had never known before: true freedom.

From that day on, Jenna lived by one truth: *Healing doesn't require control—it requires trust.* And every time she felt the urge to push, she would place her hand on her heart and whisper her mantra once more: "I release what I can't control." With each breath, her body listened—and continued its quiet, miraculous work of healing itself.

Chapter 10: Conscious Regeneration: Teaching the Body to Rebuild Itself

Regeneration is not just a biological process—it is a conscious one. Every second, over 50 million cells in your body die and are replaced. Your skin renews every month, your liver regenerates every few weeks, and even your bones completely rebuild every ten years. The human body is a living, self-repairing ecosystem. Yet when the mind is trapped in chronic stress, fear, or self-doubt, this regenerative process slows or becomes dysfunctional. Conscious regeneration means aligning your thoughts, emotions, and intentions with the body's natural blueprint for renewal.

The foundation of regeneration lies in the communication between cells. Every cell in your body is connected through biochemical and electromagnetic signals. These signals are influenced by your mental and emotional state. When you experience calm, gratitude, or love, your brain releases neuropeptides—chemical messengers that tell cells it's safe to grow, repair, and divide normally. When you live in anxiety or anger, stress hormones disrupt this communication, leading to inflammation, premature aging, and cellular exhaustion. In short, your state of mind dictates your rate of renewal.

Neuroscience and epigenetics have revealed that gene

expression—the process by which your genes create proteins and direct cell activity—is not fixed. It changes based on environment, thought, and emotion. Dr. Bruce Lipton's work on "The Biology of Belief" demonstrated that cells respond to signals from the environment, including the chemical environment created by your thoughts. By cultivating positive emotional states, practicing meditation, and visualizing health, you can literally activate genes associated with growth and repair while silencing those linked to disease and stress.

To consciously activate regeneration, the first step is creating the *environment* for healing. Cells thrive in coherence—a state where oxygen, nutrients, and energy flow freely. Deep breathing, adequate hydration, sunlight exposure, and sleep are not luxuries—they are regenerative triggers. Each of these sends a message to your body that conditions are ideal for rebuilding. You cannot regenerate in chaos. Calmness, clarity, and safety are the soil from which renewal grows.

Visualization plays a vital role in conscious regeneration. When you imagine your tissues glowing with vitality, your heart pulsing with strength, or your immune cells clearing inflammation, your brain releases corresponding neurochemicals. The subconscious mind doesn't distinguish imagination from reality—it responds to the emotional intensity behind the image. Visualize yourself not just as healed, but *renewed.* Picture your cells rebuilding perfectly, your organs functioning effortlessly, and your energy expanding each day. This mental blueprint guides your biology.

Nutrition also fuels regeneration. Foods rich in antioxidants, phytonutrients, and plant compounds—like berries, leafy greens, turmeric, and green tea—protect cells from oxidative stress and encourage the creation of new mitochondria, the powerhouses of energy production. But the *attitude* with which you eat is just as important as what you eat. When you eat with gratitude and presence, digestion improves, nutrient absorption increases, and hormonal balance stabilizes. Conscious eating becomes another form of cellular communication.

Exercise further stimulates regeneration by promoting circulation and the release of growth hormones. Movement signals to your body that it is alive, adaptable, and in need of repair. Even gentle movement like stretching or walking accelerates the delivery of oxygen and nutrients to tissues. Combined with meditation and breathwork, it becomes a powerful trifecta of physical, emotional, and energetic renewal.

Finally, gratitude and self-love act as the emotional glue that holds regeneration together. Cells respond to love the way plants respond to sunlight. The inner dialogue you maintain with yourself matters. When you think, "I am healing, I am renewing, I am grateful," your nervous system relaxes, your heart rhythm harmonizes, and your biology responds in kind. The body listens to the language of kindness more than command.

Conscious regeneration is not a mystical process—it is a return to the body's original design. You are not just a collection of organs—you are a living consciousness capable of directing your biology. When you align your breath, thought, and emotion with the truth of your body's intelligence, you awaken the ancient programming of renewal that exists in every cell. Healing then becomes more than recovery—it becomes transformation.

Daniel learns the secret

For most of his adult life, Daniel had believed that aging and decline were inevitable. At fifty, he felt older than his years—his joints ached, his energy lagged, and even his memory seemed to blur. "That's just what happens," he would say. "It's part of getting older." But deep down, he wondered if there was more to the story. He often read about people who reversed chronic illness or regained vitality through meditation, fasting, or energy work. Could the body really rebuild itself?

His turning point came during a business trip when he met an older man named Rafael. At seventy-one, Rafael moved with the agility and lightness of someone half his age. Intrigued, Daniel asked for his secret. Rafael smiled and said, "I don't fight my body—I talk to it." Seeing Daniel's confusion, he added, "Every cell in your body listens to your thoughts. I tell mine every morning that they are strong, young, and full of life. And they listen."

That conversation stayed with Daniel for months. He began researching cellular regeneration, reading about epigenetics, meditation, and neuroplasticity. The science fascinated him: the idea that the mind could turn on genes for growth and repair, or that emotion could influence cell communication. He decided to experiment on himself.

Each morning, Daniel sat quietly and visualized his body as vibrant and youthful. He pictured his cells glowing with light, his blood circulating easily, his joints moving fluidly. At first, it felt awkward—his logical mind resisted. But the more he practiced, the more natural it became. After a few weeks, he began to notice small changes. He woke up earlier without fatigue, his skin looked healthier, and the dull pain in his knees had faded.

Encouraged, he added gratitude journaling and breathwork to his daily routine. Instead of complaining about aging, he began thanking his body for its resilience. He walked barefoot on the grass, soaked in morning sunlight, and ate slowly with intention. With each meal, he silently repeated, "This nourishes my renewal." The more he respected his body, the more it responded.

After three months, friends began asking what he was doing differently. "You look younger," one said. "Did you start some new diet?" Daniel laughed. "No diet," he replied. "Just cooperation." His bloodwork confirmed what he felt—his inflammation markers were down, his testosterone and growth hormone levels had improved, and his sleep had deepened. His doctor called it "unusual for his age."

One evening, while meditating, Daniel had an unexpected realization. He saw his body not as an aging machine but as a living community of intelligent cells—each one listening, adapting, and responding to his consciousness. "I'm not in my body," he thought. "I *am* my body." The separation between mind and flesh dissolved. In that moment of unity, he felt a surge of vitality unlike anything he had known.

Years later, Daniel continued his regeneration practice, teaching others that renewal begins with awareness. He explained that the body doesn't need to be forced—it needs to be trusted. "Your cells are loyal," he told his students. "They reflect what you believe about yourself. If you believe you are breaking down, they will obey. If you believe you are rebuilding, they will obey that, too."

He never claimed to have discovered a miracle—only a relationship. The relationship between mind and matter, intention and biology. His story became a quiet testament to the truth that aging is not decline but adaptation—and that the mind can guide that adaptation toward renewal instead of deterioration.

Each morning, Daniel still wakes before sunrise, closes his eyes, and breathes deeply. He places his hand over his heart and whispers, "Every cell in my body is new, strong, and alive." As he exhales, he feels a pulse of energy move through him—warm, radiant, familiar. The conversation between his mind and his cells continues, effortless now, like an ancient rhythm remembered.

In that stillness, Daniel smiles, knowing what he once doubted: that the body doesn't age—it listens. And when you speak to it with love, it answers in the language of regeneration.

Chapter 11: The Mind of the Heart: Using Emotion as Medicine

Science once taught that the brain controls everything—that thought creates emotion. But recent discoveries have turned that understanding upside down. The heart has its own intelligence, its own nervous system, and its own form of memory. It sends more information to the brain than the brain sends to the heart. The ancient idea that emotion arises from the heart wasn't just poetic—it was biological truth. The "mind of the heart" is now recognized as a real communication center, one that can be trained to create harmony, health, and healing throughout the body.

The heart communicates with the brain and body in four ways: neurologically (through nerves), biochemically (through hormones), biophysically (through pulse and waves), and energetically (through its electromagnetic field). When emotions like love, gratitude, and compassion dominate, the heart's rhythm becomes smooth and coherent. This coherence creates balance in the nervous system, synchronizes brain waves, and optimizes immune and hormonal function. In contrast, negative emotions create erratic heart rhythms that disrupt communication and trigger stress responses. Emotion, therefore, is medicine—or poison—depending on how it is used.

The field of neurocardiology has shown that the heart's rhythmic patterns influence brain function, especially in areas related to attention, perception, intuition, and decision-making. When the heart is coherent, the brain operates more efficiently. Cognitive performance improves, creativity increases, and emotional reactivity decreases. This is why people who practice gratitude or compassion-based meditations not only feel happier—they physically function better. The body thrives on emotional coherence.

Emotions are not random; they are signals guiding the body toward balance. Anger tells you a boundary has been crossed. Fear signals potential danger. Sadness helps release attachment. But when these emotions become chronic—when anger, fear, or sorrow define your baseline—they create biochemical imbalances that suppress immunity, accelerate aging, and damage the cardiovascular system. Learning to listen to emotions, process them, and then return to the coherent emotions of love, appreciation, and peace is essential to long-term health.

To train the "mind of the heart," start by practicing *heart-focused breathing*. Close your eyes, place your hand over your heart, and slow your breath to five seconds in and five seconds out. As you breathe, imagine your breath flowing through your heart. Then evoke a regenerative emotion—gratitude, care, compassion, or love. Feel it sincerely, not just think it. Within a few minutes, your heart rhythm becomes coherent, your cortisol levels drop, and your immune markers rise. The body responds instantly to the emotional frequency of the heart.

Over time, this practice strengthens emotional resilience. Situations that once triggered stress begin to lose their power. You respond with calmness instead of reaction. The heart literally reprograms the brain. This is emotional mastery—the ability to choose coherence in the midst of chaos. Studies from the HeartMath Institute have shown that consistent heart coherence training reduces anxiety, improves focus, and increases DHEA, the hormone of vitality and longevity.

Gratitude is one of the most potent heart medicines available. It shifts your entire physiology toward healing. Gratitude tells the brain that life is abundant and safe, silencing the fight-or-flight response. It also enhances heart-rate variability, a key indicator of health and stress resistance. Likewise, compassion connects you to others on a biological level—synchronizing heart rhythms between people and amplifying collective coherence. Love, then, is not just an emotion—it's a frequency that heals.

The heart's wisdom is simple but profound: what you feel becomes what you are. Emotions are not to be controlled but cultivated. When you live with an open heart—expressing love, forgiveness, joy, and gratitude—you become a living field of healing energy. Every cell in your body tunes to that vibration. The mind of the heart becomes your true medicine, one that restores balance from the inside out and radiates healing outward into the world.

Isabel's story

Isabel had always been guided by her intellect. As a neuroscientist, she trusted logic, research, and evidence. Yet after her husband's sudden passing, none of her scientific knowledge could ease her pain. The grief felt endless—an ache that spread through her chest like a constant pressure. No amount of reasoning could touch it. Her colleagues encouraged her to take time off, but she couldn't focus. Even breathing felt heavy.

One evening, while scrolling through old research files at home, Isabel found a study she had dismissed years before—a paper on heart coherence and emotional regulation. The study described how the heart's electromagnetic field changed with emotion and how gratitude and love could restore physiological balance. In her grief, those words struck her differently. For the first time, she didn't see it as data—she saw it as hope.

That night, she decided to try the simple exercise described in the study. She sat in bed, placed her hand on her chest, and began to breathe slowly. At first, her body resisted—the sadness too thick, the ache too deep. But she kept breathing, imagining that each inhale brought light into her heart and each exhale released pain. When she finally whispered, "Thank you," she wasn't sure for what. Maybe for the years she had shared with her husband, maybe just for the strength to breathe again.

As she continued, something subtle shifted. The pressure in her chest softened. Her heartbeat slowed. She felt warmth radiating through her body, as though someone had gently placed a comforting hand on her back. For the first time in months, she slept through the night.

Over the following weeks, Isabel practiced heart-focused breathing every morning and evening. The waves of grief still came, but they no longer drowned her. She noticed that when she focused on gratitude—recalling a moment of love, a smile, a memory—her body relaxed. Her immune system, which had weakened during months of stress, began to recover. The migraines stopped. Her energy returned.

One day, while walking along a quiet trail near her home, Isabel felt an unexpected peace wash over her. She realized she didn't have to *move on* from her loss—she could carry love forward through her heart. It was as if her husband's memory had transformed from pain into light. She smiled softly, whispering, "Thank you," to the wind.

Back at the university, Isabel began integrating emotional coherence into her research. She measured participants' heart rhythms as they practiced gratitude meditation. To her amazement, the data confirmed what she had experienced personally—heart coherence improved brain function, reduced inflammation, and enhanced well-being. But more than that, she saw faces light up, eyes soften, and shoulders relax. It was science meeting the soul.

Months later, Isabel began teaching workshops titled *The Mind of the Heart*. She spoke not only as a scientist but as someone who had healed. She told her students that emotion is the body's most powerful form of communication. "Your heart," she said, "is not just a pump—it's a teacher. When you listen to it, it shows you how to heal."

During one class, a young man raised his hand and asked, "But what if my heart feels broken?" Isabel paused, placed her hand on her chest, and replied, "Then that's exactly where you begin. You breathe through it. You love through it. You let your heart teach you how strong it really is."

Years later, Isabel continued her practice. Every morning she sat quietly, breathing through her heart, feeling gratitude for the life that once broke her and then rebuilt her. She no longer sought healing from the outside; she lived it from within.

Her story became a living demonstration of her own research: that emotion, when felt consciously and with love, is the purest medicine the body can create. The mind may understand healing, but the heart *becomes* it. And through that realization, Isabel finally understood what every heartbeat had been trying to tell her all along—that love, when truly felt, never ends.

Chapter 12: Awakening Inner Intelligence: The Brain–Body Symphony of Complete Healing

Within you exists a silent orchestra of intelligence—billions of neurons, trillions of cells, and countless molecular signals all working in concert to sustain life. This intelligence is not random; it is self-organizing, adaptive, and responsive to consciousness itself. Every heartbeat, breath, and chemical reaction happens with perfect precision, guided by the body's inner wisdom. Awakening this intelligence is the final step in the healing journey. It is when awareness, trust, and intention merge—transforming the body from a system of survival into a symphony of coherence and creation.

Inner intelligence operates through communication. Your brain and body are in constant dialogue through electrical impulses, hormones, peptides, and electromagnetic signals. Thoughts become chemistry; emotions become electrical energy; beliefs become physical structure. When this communication is clear, health flourishes. When it's distorted—through fear, doubt, or disconnection—disease begins. Healing, therefore, is not about fixing the body, but restoring the conversation between the brain, the heart, and the cells.

Modern neuroscience and quantum biology now suggest that consciousness itself may be the ultimate organizing force behind this inner communication. Your cells are not passive—they listen. They adapt to the energy of your attention. When you focus with fear, they tighten, conserve, and defend. When you focus with love, they expand, regenerate, and create. This is the essence of awakened healing—where awareness becomes the conductor, and every system of the body follows its rhythm.

To awaken inner intelligence, start by tuning in. Spend a few minutes each morning in silence, scanning your body from head to toe. Notice areas of tension, warmth, or energy. Instead of judging them, simply observe. This act of mindful attention creates a bridge between the conscious and subconscious mind. You begin to "hear" your body's language again—the subtle sensations that guide you toward what it needs. Over time, intuition sharpens. You will know when to rest, when to move, what to eat, and what to release.

The next step is harmonization. Through breath, emotion, and thought, align your internal systems into coherence. Deep, rhythmic breathing signals safety to the nervous system. Positive emotions like gratitude and love generate heart coherence, which synchronizes the brain's electrical activity. Focused thought gives the body a clear intention. When all three align—breath, emotion, and thought—the brain and body operate in unison. Studies show that this synchronized state optimizes immunity, hormonal balance, and even gene expression.

Movement can also awaken inner intelligence. Practices like yoga, tai chi, and dance reconnect the mind to the physical self. Every stretch and movement becomes a dialogue with your body's wisdom. You begin to sense that the body is not a machine but a conscious partner. Listening to its rhythm teaches patience, presence, and compassion.

Finally, surrender is the key to letting inner intelligence take the lead. Once you've set the intention for healing, release it into trust. The conscious mind sets the goal, but the subconscious and the body execute it. Trying to control the process creates tension; allowing it invites flow. This is where healing deepens from mechanical to miraculous.

You are not separate from the intelligence that created you. The same consciousness that formed galaxies orchestrates your heartbeat and breath. When you quiet the noise of fear and tune into the deeper current within, that intelligence awakens fully. Healing becomes effortless—not because you forced it, but because you remembered how to listen.

Awakening inner intelligence is the realization that your body is not your limitation—it is your ally. It holds a wisdom older than memory, waiting to respond to your awareness. When mind, heart, and body move as one, healing ceases to be a destination and becomes your natural state.

Amara's Brain–Body Symphony

Amara had always been logical. She followed instructions, took her medications on schedule, and did everything her doctors advised. But after years of chronic fatigue and recurring illness, she felt trapped in a body that refused to cooperate. "I'm doing everything right," she told her physician, "so why am I still sick?" He had no clear answer. Her blood tests looked fine. Her scans showed nothing abnormal. Yet deep down, Amara felt something vital was disconnected—not in her body, but between her body and mind.

A friend suggested she try a body-awareness retreat focused on "awakening inner intelligence." Skeptical but desperate, she agreed. On the first morning, the instructor—a serene woman with kind eyes—asked the group to sit in silence and "listen to the body." For ten minutes, they did nothing but breathe and observe. Amara's mind protested. *This can't possibly help,* she thought. But then, she felt something unexpected: a pulsing warmth in her abdomen and a soft vibration in her chest. It wasn't pain—it was presence.

The instructor explained that this awareness was her body's intelligence communicating. "You don't have to control healing," she said. "You have to hear it." That sentence stayed with Amara. For the first time, she began to see her body not as an obstacle but as a messenger.

Over the next few days, Amara practiced tuning in. Each morning, she placed her hands over her heart and asked silently, *What do you need today?* Sometimes she felt a craving for rest, sometimes the urge to move. Sometimes she felt emotions surface—old sadness, fear, even guilt. Instead of resisting, she breathed through them, letting her body guide the release.

During one session, the group practiced heart–brain coherence breathing. As Amara focused on gratitude, she felt a subtle wave rise from her heart to her head—a warmth that cleared her mind and brought a surprising calm. Later, the instructor explained that her nervous system had synchronized, allowing her body's healing mechanisms to activate. "This," she said, "is the language of inner intelligence."

When Amara returned home, she made listening her new routine. Before meals, she asked her body what foods felt right. Before sleep, she thanked it for carrying her through the day. When tension appeared, she breathed into it instead of ignoring it. Over time, her energy increased, her digestion improved, and her chronic fatigue faded. Her doctors couldn't explain it, but Amara didn't need them to. She understood: her body had been waiting for her to participate, not command.

Months later, she began teaching mindfulness and self-healing techniques to others. In her workshops, she told participants, "Your body knows how to heal—it's been doing it since the day you were born. All you have to do is stop shouting at it and start listening." She demonstrated how simple breathing and gratitude practices could create measurable calm. Watching others find peace reminded her of how she began—confused, tired, disconnected—and how quickly everything changed once she stopped trying to control and started to trust.

One evening, after leading a session, Amara sat quietly in the darkened studio. The room was still, her breath steady. She placed her hand over her heart and smiled. That soft vibration she first felt at the retreat was now familiar—a gentle rhythm of harmony between mind and body. She realized that healing had never been about effort; it had always been about remembrance. Her body wasn't broken—it had been waiting for her awareness.

In that moment, she felt profound gratitude—not only for her recovery but for the partnership she had rediscovered. Her body had become her teacher, her consciousness the student. Together, they formed a symphony of intelligence that no external cure could replicate.

From then on, Amara lived differently. She worked with presence, rested with intention, and spoke to her body with kindness. Whenever someone asked how she healed, she would say softly, "I didn't heal myself. I listened, and my body showed me how."

Her story became a testament to the quiet truth that healing does not come from outside-in but from inside-out—from the intelligence that has always been there, waiting to be awakened. And in that awakening, Amara found not only health, but harmony—a living proof that the greatest healer we'll ever meet already lives within.

Epilogue: The Return to Wholeness: Remembering What the Body Always Knew

Healing is not a destination. It is not an achievement, nor is it a miracle reserved for a few. Healing is a remembering—the slow and sacred return to the truth that your body has always known how to thrive. Each heartbeat, each breath, each quiet pulse of energy within you is proof that life never stopped fighting for you. What changed through this journey was not the body, but the awareness through which you began to see it. You discovered that health is not created through force but through harmony; not through fear but through trust; not through separation but through union. The mind and the body, once divided by misunderstanding, found their way back to one another—and through that reunion, you became whole again.

When you began this exploration of the brain–immune connection, you may have believed that healing was purely physical, that it depended on medicine, supplements, or systems outside yourself. But each chapter revealed a deeper truth—that the power to heal does not reside in what you take, but in what you activate. You learned that your brain speaks to your immune system through emotion, that your thoughts shape your biology, and that every feeling, belief, and word you carry sends a signal that either nourishes or depletes your body's strength. What you discovered was not a theory—it was a language. The language of coherence. The silent conversation between neurons and heartbeats, between breath and chemistry, between the unseen and the tangible. You began to remember that your health was never separate from your consciousness—it was its reflection.

As you practiced awareness, breath, and emotional balance, your inner world began to shift. The chaos quieted. The noise of constant worry started to dissolve into stillness. You realized that peace is not an emotion—it is a frequency that heals. In that peace, your cells listened and responded. The inflammation began to ease, sleep deepened, and energy returned. You saw that surrender is not about giving up control, but about giving space for the body's intelligence to work. You saw how gratitude rewired your heart, how visualization reprogrammed your brain, and how faith turned chemistry into transformation. What once felt mystical became measurable; what once seemed invisible became undeniable. Each moment of trust, each breath of love, was an act of medicine.

Healing, you now understand, is not the absence of illness—it is the presence of alignment. You can still face challenges, setbacks, or pain, but now you meet them differently. Instead of fighting your body, you listen to it. Instead of silencing your emotions, you let them speak. You know now that your body is never betraying you—it is always communicating. A symptom is not punishment; it is information. A moment of fatigue, a wave of anxiety, a tightening in the chest—all of it is your body's language reminding you that something needs attention, compassion, or rest. When you respond to those messages with love instead of fear, the body responds in kind. The war ends, and the partnership begins.

There will be moments in life when stress returns, when chaos intrudes, when the world outside becomes loud again. But you now carry the tools to return to stillness—to the healing state that lives within you. A few slow breaths, a hand over the heart, a single thought of gratitude—these are your gateways back to balance. You've learned that regeneration, immunity, and energy are not mechanical processes but responses to awareness. The body listens to the emotional tone of your consciousness more than it listens to any prescription. When you speak to it with kindness, it obeys with vitality. When you trust it, it responds with strength. When you love it, it transforms.

Perhaps the greatest revelation of all is that you were never just healing your body—you were awakening yourself. Each step along this path taught you that you are not a collection of organs but a field of intelligence, vibrating in harmony with the universe that created you. The same intelligence that spins galaxies lives in your cells, orchestrating a symphony of biochemical miracles each second. Every time you breathe consciously, you are aligning with that intelligence. Every time you choose love over fear, you are syncing with the frequency of life itself. You have discovered that healing is not something that happens *to* you—it happens *through* you. You are not separate from the medicine; you are the medicine.

In this awakening, you begin to see that true health extends beyond the physical. It is emotional stability, mental clarity, spiritual peace, and energetic coherence all moving together as one. You are no longer a patient waiting for recovery—you are a participant in the creation of wellness. You are an artist painting balance with your thoughts, sculpting vitality with your emotions, and composing harmony with your breath. Every act of mindfulness is a brushstroke. Every expression of gratitude is a color. Every moment of love is the light that gives your life its radiance. The masterpiece is you—whole, aware, and alive.

And so, as this book closes, your journey does not end—it deepens. The science, the stories, the practices—all of them were signposts leading you back to a truth that is both ancient and eternal: that the mind, the heart, and the body are not separate forces, but facets of the same consciousness. You are both the observer and the observed, the healer and the healed. The intelligence that beats your heart knows how to restore every part of you—it always has. Your only task is to remember, to listen, and to stay aligned with that knowing. The body you live in is not your limitation; it is your instrument. And when you play it in harmony with awareness, love, and trust, you awaken a music so pure that it heals everything it touches.

The return to wholeness is not about becoming someone new—it is about remembering who you've always been: a being of light, consciousness, and limitless potential, capable of turning thought into biology, emotion into energy, and awareness into health. As you move forward, carry this truth with you. You are not broken. You are not waiting to be healed. You are already whole—and the more you remember that, the more your body will too. Each breath, each thought, each act of compassion is a reminder to the universe of your alignment with life itself. And in that alignment, healing never ends—it becomes your way of being.

So, breathe deeply. Smile gently. Trust completely. Your body knows the way home. All it ever needed was your awareness to walk beside it. And now that you have returned to that partnership, the journey of healing will continue endlessly—effortless, graceful, and alive. For in the silence of every heartbeat and the stillness of every breath lies the same eternal truth whispered through every cell of your being: you are life healing itself.

www.ingramcontent.com/pod-product-compliance
Lightning Source LLC
LaVergne TN
LVHW020237210926
848481LV00016B/2348